SOTHIS PRESS

Sothis Press aims to completely reimagine the public domain book with an emphasis on universal access to captivatingly designed print editions. We believe that technology will enable every single book ever published to be perpetually accessible in beautiful print form. Sothis is beauty in ubiquity.

www.sothis.press
1 Waterhouse Square, London, ECIN 2ST

In Litchfield Hills

An Illustrated Work of Litchfield County,
in Which the Picturesque Features of Each
Town in the County Are Set Forth

George Alson Marvin

SOTHIS PRESS

ISBN 978-1-0368-0566-1

IN LITCHFIELD HILLS.

AN ILLUSTRATED WORK OF LITCHFIELD COUNTY,

In Which the Picturesque Features of Each Town in the County are Set Forth.

PUBLISHED AND EDITED BY

GEORGE ALSON MARVIN.

HARTFORD, CONN.
R. S. PECK & CO., PRINTERS,
1897.

CONTENTS.

	Page.
In Litchfield Hills,	10
Litchfield,	12
Winchester,	19
Torrington,	30
North Canaan,	33
Norfolk,	39
Salisbury,	55
Thomaston,	64
New Milford,	66
Sharon,	69
New Hartford,	77
Watertown,	81
Woodbury,	90
Roxbury,	94
Harwinton,	97
Plymouth,	99
Colebrook,	103
Barkhampsted,	108
Bethlehem,	110
Cornwall,	111
Kent,	115
Goshen,	118
Canaan,	121
Warren,	125
Morris,	127
Bridgewater,	131
Washington,	132

Preface.

In the following pages the author has endeavored to write a clear and brief description of the picturesque in each town in the county.

"In Litchfield Hills" is not to be considered a historical work, although many interesting points of history are touched upon.

As nothing of this character on our county has been placed before the public it is the object of the author to place upon the market a pretty souvenir of "Our Mountain County" at a price within the reach of every one.

The writer has personally visited every town and collected the necessary material besides selecting the subjects for illustration.

Although the work has involved a large amount of labor and there have been many difficulties to overcome the task has been most fascinating.

Much is due to those who have taken an interest in the preparation of this volume, and have aided by valuable suggestions. Many contributions have been made by certain of Litchfield's brightest literary minds which have added much to the value of the work.

G. A. M.

NORTH STREET, LITCHFIELD.

IN LITCHFIELD HILLS.

Before taking up in detail each town in Litchfield county perhaps it would be of interest to take a cursory glance at the "Mountain County of Connecticut." This northwestern corner of our commonwealth is one of the most picturesque and delightful spots in all New England.

Over a thousand square miles of pleasing diversity of scenery interspersed with lakes, mountains and valley sprinkled with quaint hamlets and busy villages, while many places have been beautified by the hand of man, make this section one of the favorite portions of Nature's great play ground for the tired denizens of our large cities.

The geologist could spend years here studying the formation and history of these rugged hills and beautiful valleys. In Salisbury are the inexhaustible ore beds which have been worked for over a century, while the lover of Nature will find rugged mountains and lakes scattered in prodigal profusion throughout the town. Further east are the beautiful meadows and limestone formations of Canaan. Then comes a ridge of Archæan rock beginning at Norfolk and stretching diagonally across the county with the oldest and hardest portion between Norfolk and Winsted.

Besides the geologist, hundreds view the wonderful "Tipping Rock" in North Goshen, which was borne to its present resting place thousands of years ago during the glacial period.

In the southern part of the county are also many geological features. Here are some of the finest granite quarries in the country, peculiar iron formations, and garnet mines.

The artist who desires to sketch or paint from Nature will find in Litchfield a perfect mine. During any season of the year picturesque bits of scenery, all varied and beautiful, may be found in every town in the county, but the artist especially loves to paint among the beautiful Norfolk or rugged Cornwall hills.

The Litchfield hills were recently described as a region of lakes, mountains, torrents, glens, lovers' lanes, rocks, and echoes, and all are bathed in romance and legends.

Down in the picturesque old towns in Southern Litchfield the Red Man warred and wooed, from the hill tops burned the signal fire calling the tribes to deadly foray; and finally after the last war-whoop had been sounded, and the noble Red Man lay stretched upon mother earth, he was laid to rest beside some beautiful water fall, or in some fertile meadow beside a peaceful brook and the story has come down to us clothed in mystery and romance.

Litchfield county is rich in historic interest and every inch is hallowed ground.

In the township of Litchfield the famous Wolcott and Marvin families melted the leaden equestrian statue of George III, which had been brought from Bowling Green, New York, into bullets, which were used to repel the hired invaders of the tyrant king.

Time and space forbid us naming or giving a record of the distinguished men who were born or lived in this county, United States senators, congressmen, governors, chief justices, clergymen, poets and authors, all swell the list of those who have not only been leaders of thought and action in the formative period of our country's history—but many are now illustrious leaders in this busy age of progress and competition.

LITCHFIELD.

For years the writer has known Litchfield by report as a place rich in historic interest, as a town of broad streets lined with fine old trees, and as a resting place for hundreds of tired dwellers of our large commercial centers.

But the writer had formed no adequate idea of this beautiful old town until he was obliged to go there when he was engaged upon the work of preparing "IN LITCHFIELD HILLS."

WELL ON HENRY WARD BEECHER PLACE.

Litchfield does not lie in a valley but sits upon a hill or plateau 1,114 feet above the sea.

From under the noble elms of this town one looks out over the surrounding country upon most diversified scenery; down the picturesque Shepaug Valley or northward to Ivy mountain tower which seems to guard the peaceful Goshen Hills or he may catch glimpses of shimmering lake or look over cultivated fields and miles of beautiful forest. Truly "Dame Nature" has been very kind to Litchfield.

NORTH STREET.

One of the chief ornaments of this town are the broad streets, level and lined with magnificent elms, which extend throughout the town. In fact many of these trees have a history. Two large elms in front of J. L. Judd's residence on West Street were planted by South Carolina's distinguished statesman, John C. Calhoun, when pursuing his studies under Judge Reeve in 1805.

One of the finest elms in the village stands on the corner of North and West Streets and in years gone by served as a whipping post.

It is said that a large willow which stands near Mrs. W. C. Noyes' grew from a riding stick, which her grandfather,

BIRTHPLACE OF HENRY WARD BEECHER.

Col. Benjamin Talmadge, who was a member of Washington's staff during the Revolution, stuck in the ground.

In this town it seems as if the past and present linked hands, for there is no town of its size which has so many houses built previous to 1800 and are still occupied by descendants of the original owners.

Notable among these is the old Wolcott house where lived Oliver Wolcott, signer of the Declaration of Independence. The old Beecher homestead where lived the Reverend Lyman

Beecher, D. D. and where Henry Ward and Harriet Beecher Stowe passed their childhood, is a point of interest to visitors.

In this, our shire town, stands the stone court house upon the site of an old fort, which was built to protect the small settlement of whites from their foe, the Red Man.

Litchfield is noted for its lakes, and no one should leave town without viewing Bantam Lake, the largest body of water in the state, covering over 1,200 acres.

FIRE DEPARTMENT BUILDING.

The residents of Litchfield take great pride in the appearance of their village and the result is acres of well-kept lawns and over five miles of concrete sidewalk, for though Litchfield is a borough, its population is but a little over 5,000.

The beauty of the scenery and the environments of the town have aided in bringing hundreds of summer guests to Litchfield. The result has been the building of many beautiful summer residences. Notable among these is Captain Van Winkle's new house. We quote the following from the "Connecticut Quarterly":

"Go to Prospect Hill, a little northwest of the village, and look for many miles to the east, south and west. Almost below you is the lovely village of Litchfield—a jewel, clean and white, in the setting of leafy green. Beyond, and higher is Chestnut Hill—a picture of highly cultivated farms in a frame of blue sky and sunshine; then Morris hills to the south, beyond the placid lake sparkling in the sunbeams; while to the southwest Big Mount Tom lifts his high head as unconcernedly as when two hundred years ago the lurid flames upon his bare pate called the Indians to their chieftain's side for deadly foray. Then to the west, is Bare Hill, the pinnacle overlooking

CASINO.

Waramaug Lake, the blue lines of Quaker Hill; then Mount Prospect, with its stores of mineral wealth, then the misty outlines of the Catskills, Mohawk Mountain with its disabled tower. Such is the vista that those who are fortunate enough to live in Capt. Van Winkle's house can see each day,—an almost boundless view of Nature, everywhere improved by art."

The influx of summer guests has led some of the enter-

PROF HOPIN'S RESIDENCE.

prising citizens of Litchfield to erect the Casino. In connection with the Casino are six tennis courts and golf links.

Through the generosity of one of Litchfield's citizens, a Fire Department Building (perhaps one of the finest in the United States has been erected.)

THE HAWK HURST.

The building which is built of stone and brick contains besides Fire Department Home, bowling alleys, billiard and pool tables, reading room, and all accessories.

Litchfield has several fine church buildings. The town is peculiarly favored with hotels. The Hawk Hurst, a fashionable summer hotel about four minutes walk from the station, is situated on the brow of a hill and directly opposite the Casino. Broad piazzas surround this home-like hotel from which one may obtain a view of hill, valley and lake.

The United States Hotel was built on the site of the original building which was erected by John Phelps, the ancestor of the Hon. E. J. Phelps, ex-minister to Great Britain.

The graceful elms, the beautiful scenery, fine buildings and historic old homes shrouded in romance and legends, together with the hospitality of the many fine old families have made the name of Litchfield known far and near.

If there is any one that doubts that Winchester lacks the picturesque features which are so characteristic of Litchfield County, let him but mount a wheel at Colebrook Station and go pedalling down the fine road which leads to the Borough of Winsted.

MEMORIAL TOWER.

Much of the distance is lined by trees while in some places, as the rider glides along the road he may look down the precipitous sides of a deep gorge, through which flows the turbulent Mad River which at last is made to turn many water-wheels in the busy Borough of Winsted.

In order to get a fair idea of Winsted one should climb one of the many eminences which surround the crescent shaped town. Perhaps one of the most accessible points is the Memorial Monument which stands but a short distance from Main street. This monument, which was erected a few years since by the citizens of Winchester, is built of granite and from foundation to top measures

MAD RIVER.

GILBERT HOME, WEST WINSTED

seventy feet. As one ascends the stairs he will notice that the interior is covered with marble tablets on which are graven the names of those who fought for the preservation of the Union.

GILBERT SCHOOL, EAST WINSTED.

Many fortunes have been made in this town by enterprising business men in the days gone by, but probably none have been put to better use than that which was accumulated by the late William L. Gilbert.

EAST WINSTED PARK

RESIDENCE OF EUGENE POTTER.

On a picturesque site a short distance from the Memorial Monument is located the endowed Gilbert Home for the orphans of Litchfield county.

RESIDENCE OF GOV. LORIN COOKE.

WAKEFIELD BOULEVARD.

SIESTA LODGE, HIGHLAND LAKE.

In East Winsted we find a handsome structure known as the Gilbert School. This school, like the home, is endowed. Its object is to prepare for the leading American colleges, and is free to those residing in the town of Winchester. Directly opposite the Gilbert School lies Winsted's fine old park with many shade trees and an ornamental fountain.

Churches of nearly all denominations are found in the town, while the Y. M. C. A. is one of the strongest organizations of its kind in the county. Winsted too, is fortunate in possessing the Beardsley Library.

The visitor who would spend a day in this town is sure of finding first-class hotel accommodations, for the hotels are ranked among the best in the Naugatuck Valley. Winsted has many fine streets lined with handsome residences, the most expensive and picturesque being the Potter Mansion.

Just at present Winsted is very much before the public eye for it is the home of Governor Cooke. The Governor has

recently completed the building of a pretty residence on Monroe street.

With all her advantages, the Borough of Winsted is most easy of access. The Philadelphia, Reading and New England Railroad runs directly through the town, while the Naugatuck division of the New York, New Haven and Hartford Railroad has its terminal in the heart of the borough.

Less than half a mile from Main street and at an elevation of two hundred feet above the town lies the beautiful Highland Lake. This expanse of water which extends three miles, has an area of nearly five hundred acres, and is encircled by forests and sloping meadows, while in some places rugged bluffs reach down almost to the water's edge. Through the generosity of the late Harvey Wakefield, a fine drive known as the Wakefield Boulevard extends around the lake, a distance of seven miles. Already several have built cottages on the shores of the lake and spend their summers there. On moonlight summer evenings the pink-a-punk of the banjo or some jolly boating song is wafted across the silvery waters as some gay party comes rowing across the lake.

But this Highland Lake is not the only sheet of water within the township of Winchester, for two hundred feet higher and within less than a mile, snugly ensconced in the rugged mountains lies Crystal Lake, whose outlet is a stream which flows into Highland Lake. Several miles north of Winsted an elaborate system of artificial canals and lakes have been constructed into which a portion of the waters of Mad River have been turned, whence it is led through the Gilbert tunnel, bored for more than a half a mile through the solid rock of a mountain where it at last rushes into Crystal Lake, so that despite the most prolonged drouth, Winsted has stored up an abundant motive power for her many factories.

Only a mile west of Winsted's famous boulevard lies Platt Mountain, the highest cultivated land in the state. A tower fifty feet in height has recently been erected on the mountain and from its top one may obtain a sweeping view over four counties. Let us leave the reader at this point viewing the diversified scenery of Northwestern Connecticut.

HIGHLAND LAKE, 489 ACRES

TORRINGTON.

Torrington, the busiest borough in the county, is picturesquely situated south of Winsted surrounded by high hills.

MIGEON AVENUE.

The village is beautifully laid out, the streets being broad and straight.

Migeon Avenue is probably the handsomest street, for it is lined with graceful elms which form a perfect arch in many places. Litchfield street is another of Torrington's pretty thoroughfares.

Hundreds of pretty residences are situated within the limits of the borough, while handsome churches of various denominations are located in the central part of the town.

Every resident takes pride in the handsome Y. M. C. A. building which is located on Main street. This building, which is built of brick, contains a reading room, parlor, recitation rooms, office of the secretary, and a gymnasium thoroughly fitted with modern appliances.

BIRTHPLACE OF JOHN BROWN

A very fine view of the village may be obtained from Torringford street To the north lie the blue Goshen hills

GLEN NEAR JOHN BROWN PLACE.

while in the foreground nestled in the wooded hills lies the busy little village, its tall chimneys breathing forth dark smoke wreaths.

Here are located the Coe Brass Works, one of the largest concerns of its kind in the country, while various other mills which furnish employment for hundreds of operatives are situated in different parts of the town. It is on this street that Charles Battell Loomis, the humorist, makes his home.

Torrington has also played an honorable part in her country's history. It was in this town the first anti-slavery convention of Litchfield county was held. And here was the birth-place of John Brown. The house in which this hero was born is located about three miles from Torrington on the

LITCHFIELD STREET.

Goshen road, and is still standing, at present being occupied by a family belonging to the race for whom this brave old man laid down his life.

Near the old John Brown house is a very pretty glen which would furnish a fit subject for an artist. In fact, one can hardly go anywhere in this pretty township without finding the picturesque.

NORTH CANAAN.

Canaan has in many respects a unique situation. Lying in the beautiful Housatonic Valley at the junction of the Berkshire division of the New York, New Haven & Hartford Railroad and the New England division of the Philadelphia & Reading Railroad it is one of the most accessible towns in Litchfield county.

Although its elevation is such as to iusure pure air and sweeping views, many portions of the town are comparatively

OLD ELM.

level so that there is room to lay out broad avenues should it ever become desirable.

Although the scenery is not as rugged and pronounced as that of some sister towns, one may feast his eyes upon the blue Taconic range or follow the wanderings of the "bonny Brom-

MAIN STREET.

foxit," better known as the Blackberry River, as it comes bu bling and foaming down from the Norfolk hills to wind in a out among the peaceful meadows of Canaan, or he may wa along the banks of the broad and romantic Housatonic Riv

Many times has the writer stood on the rear platform o west bound train on the Philadelphia, Reading & New Engla Railroad on a beautiful autumnal afternoon and watched t wild and romantic scenery as the train rushed on its w through deep rock cuts or along the side of some high ele

BROMFOXIT.

tion of land. The view indeed is magnificent as one lo down upon the picturesque valley of the Blackberry River, views the rugged mountains clothed with hardy growths wood, the foliage of which is bright scarlet or shimmer gold.

As the visitor walks down Main street he is at once stri by the number of pretty residences surrounded by well k lawns, (the property of Canaan's prosperous citizens), wh line both sides of the street. Canaan too has many fine hor more than a century old, every one of which is shrouded romance, and each has its own legend.

MARBLE QUARRY, EAST CANAAN

Canaan's big elm a short distance north of the village is the Mecca of many visitors.

Among the public buildings of this town is a cosy little library which contains the Douglas Library, consisting of three thousand volumes.

Canaan also has four churches.

For many years this town has had a fine water plant, the water being obtained from a series of inexhaustible springs a little more than a mile from the village.

Connecticut's beautiful State House at Hartford is admired by all who visit that historic old city, and it is a notable fact that the marble for this stately structure was obtained from the quarry at East Canaan.

For over half a century a fine quality of lime has been shipped from this town to various parts of the state.

Down on the picturesque lower road are located the two blast furnaces of the Barnum, Richardson Company, always busy reducing ore, the daily output being nearly forty tons.

Although there are no large factories in the town the fine water power of the Blackberry river is most excellent and would seem a most pleasing inducement to the manufacturer who would avoid the expense of fuel and engines.

Truly Canaan has a future.

NORFOLK, FROM CRISSEY HILL.

NORFOLK.

No township within the borders of Litchfield County has a greater reputation for the picturesque and beautiful than the grand old town of Norfolk.

This beautiful village ensconsed among the blue Litchfield hills is the highest point of land reached by rail in the state, and is known as a popular summer resort from New York

SIGN POST NEAR PARK

Harbor to the Golden Gate, and from Canada to the Gulf of Mexico, while its elevation of over 1,300 feet above tide water insures pure air and cool mountain breezes.

LAKE WANGUM.

In this pretty little town the seeker after health or pleasure may find acres of velvet lawns which compare favorably with the trim lawns of Newport, or he may climb rugged mountain paths if he be a pedestrian, and obtain sweeping views of Litchfield and Berkshire Hills, or if he be an ardent follower of Izack Walton he may paddle his canoe over the placid waters of Lakes Wangum, Doolittle or Toby, and

LIBRARY.

if he be patient he may carry away some very fine strings of bass and pickerel.

Although Nature has been very lavish the town owes much of its adornment to the Eldridge and Battell families.

On the summit of Haystack, which has an elevation of 1620 feet above the level of the sea, and is situated about a mile in a northwesterly direction from the village is a belvedere, which was erected several years ago by the direction of the late Hon. Robbins Battell.

The Norfolk Library, built in 1888, is considered one of the finest in the state, and contains a reading room which is supplied with over forty of the leading dailies and periodicals, a conversational room, and a library proper which contains

ROOM, LIBRARY

HALL OF LIBRARY.

over 9,000 volumes. The building also contains apartments for the librarian.

This building, which was designed by Keller, was erected by Miss Isabelle Eldridge in memory of her father and mother, Joseph and Sarah Eldridge.

The residents of the town and visitors also have free use of the Eldridge Gymnasium, which is considered one of the most complete institutions of its kind in the state, and is fully

ELDRIDGE GYMNASIUM.

equipped with modern and approved apparatus and appliances, the whole being under the constant supervision of a competent director. He who desires an extensive view with little exertion will find no better spot than the large overhanging veranda of this building.

From this luxuriously furnished piazza one may obtain a view of the beautiful Canaan Valley with Mt. Everett in the background. Another attraction of this popular resort is the tennis courts connected with the gymnasium.

The visitor should not neglect to spend a brief time loitering about Norfolk's historic old park, or admiring the

BUTTERMILK FALLS.

beautiful fountain, designed by Sanford White, the bronze Gaudens, which stands at the lower point of the triang park. This fountain was erected in the fall of 1889 by M Mary Eldridge in memory of the late Joseph Battell. Welch memorial fountain, another of Norfolk's gifts,

VERANDA GYMNASIUM.

erected in 1895 by the Welch family in memory of the late William Welch, and is placed at the junction of Main and M streets.

Ever since Norfolk's first pastor, Rev. Ammi Robbi entered upon his pastorate in the fall of 1761, Norfolk has be known as an educational center.

Some ten or twelve years ago a model public scho building was erected. In 1884 Miss Anna Battell in conne tion with her brother, Hon. Robbins Battell (lineal descendan of Rev. Ammi Robbins) established the Robbins School, t first permanent preparatory school of northwestern Connectic

Recently a fine athletic field, the gift of Mrs. Carl Stoeck has been added to the school campus. West of the pa

CAMPBELL FALLS.

stands a typical New England church with a Wren steeple. Another gift is the chime of bells which have been placed in the old church tower and whose sweet cadence may be heard every quarter hour.

Beside the Congregational church, and facing the park is a chapel built by Mrs. Urania Humphrey in memory of her father and mother. This building, which was designed by Cady, is of granite and contains a lecture room and parlor.

The town also has a pretty Roman Catholic church, a quaint old Methodist church and a new Episcopal church, which is supported largely by the summer residents.

VIEW ON LAKE DOOLITTLE.

The Hillhurst, a large airy hostelry accommodating over a hundred guests, is situated a half a mile in an easterly direction from the village green. From the spacious verandas of this summer hotel beautiful and extensive views may be obtained.

Fairlawn, a pretty cottage on Maple Avenue accommodates about twenty of Norfolk's summer guests.

ROBBINS SCHOOL.

The Stevens House, a commodious hotel surrounded by a beautiful lawn and shade trees, is open throughout the year.

Although Norfolk is not noted for its manufacturing, a few enterprising firms have located on the banks of the Blackberry River, a picturesque little stream which drains nearly twenty miles of hilly country. In one place this stream has a fall of 125 feet in less than a mile.

One of the many fine bits of scenery about Norfolk is formed by this river as it plunges in a series of connected

STEVENS HOUSF.

tumbles over the rocks west of the mill dam, forming the Butter Milk Falls.

Norfolk no longer depends upon the "old oaken bucket" for her water supply. Some two years ago work was commenced on Norfolk's new water plant. The supply is obtained from Lake Wangum, a beautiful sheet of water located about four miles from the village on the top of a high mountain, and encircled by forests. Its height of nearly 250 feet above the village green, the fact that there are but two or three buildings located on the entire water shed, and that the lake is supplied by never-failing springs ensures perfect freedom from contamination.

HILLHURST.

It has been said that Norfolk's drives would require a volume, but in a work of this kind space will permit of but mentioning a few.

Campbell's Falls, located about four miles from Norfolk is one of the many pretty drives. Our brief description we quote from an article published in the "Norfolk Chimes."

"The best scenery can be had on this trip by taking the road over Ball Mountain and returning by way of Canaan valley. Leaving the wagon on the roadside one descends by a

A NORFOLK RESIDENCE.

steep path thickly strewn with pine-needles into a grass-grown and wooded ravine, and there for the first time one catches a glimpse of the Falls. From a height on the opposite side of the ravine the sparkling water dashes over the mossy rocks, sometimes falling into little rocky nooks where it forms a secluded lake until the next heavy rain forces it out and on its way. Bounding over the rocks it at length flows into a bubbling brook which winds in and out among the trees till it is lost to sight. By following this brook now walking along the bank, now skipping from stone to stone, one sees many beautiful ferns, mosses, and flowers, which add greatly to the

SUMMER RESIDENCE OF PROF. GOODNOW. SOLDIERS' MONUMENT. VIEW OF VERANDA AT TAMARACK LODGE.

WELCH MEMORIAL FOUNTAIN. TOBY POND.

beauty of the place. Along the sides of the falls are large rocks and boulders by the aid of which, with careful climbing, one may reach the top where the water makes its first bold dash over the rocks. From this spot the view of the ravine is very charming, although most of the falls are hidden from sight."

Far View, about two miles west of the village, is a popular drive, while he who prefers a steep climb to a drive, may invade Lovers' Lane. A much frequented drive is to Moses Hill, situated two and a half miles south of the village on the Litchfield Road. From this eminence, which is owned by Dr. Fred-

STONE HOUSE, SUMMER RESIDENCE OF LIEU.-GOV. DEWELL, W. NORFOLK.

erick S. Dennis of New York, one may look over the rolling hills across the state of Massachusetts and see the Green Mountains of Vermont; in fact, nothing obstructs the view in any direction for miles and miles around. A drive of two miles below Moses Hill takes us to South Norfolk by the silent old water wheel and its picturesque surroundings.

Another of the attractive features of Norfolk are the handsome summer residences, which are enclosed in well-kept grounds. The finest of these is the recently completed residence, built of dove colored marble, of Mr. H. H. Bridgman of New York.

Tamarack Lodge, the property of Dr. Frederick S. Dennis, is also considered one of Norfolk's most beautiful summer homes. Another residence that attracts much attention is owned by Professor Frank Goodnow of Columbia College. The "Stone House" owned by Lieutenant-Governor Dewell of New Haven, is also an object of much interest.

Besides these summer homes there are many fine old homesteads, some over a century old, in the town. Space forbids us to mention the many historical associations connected with the town.

He who would enjoy an ideal summer, the autumnal beauties or pleasures of a New England town, will find no pleasanter place in all New England in which to cast his lot, than Norfolk.

SALISBURY.

Salisbury, the largest township in the county, presents almost every variety of scenery. Rugged mountains lift their hoary heads toward the clouds, while at their base shimmering sheets of blue water splash upon a sandy beach; tiny rivulets come bubbling down from the hillsides to add their mite to the broad bosom of Connecticut's most picturesque river, the

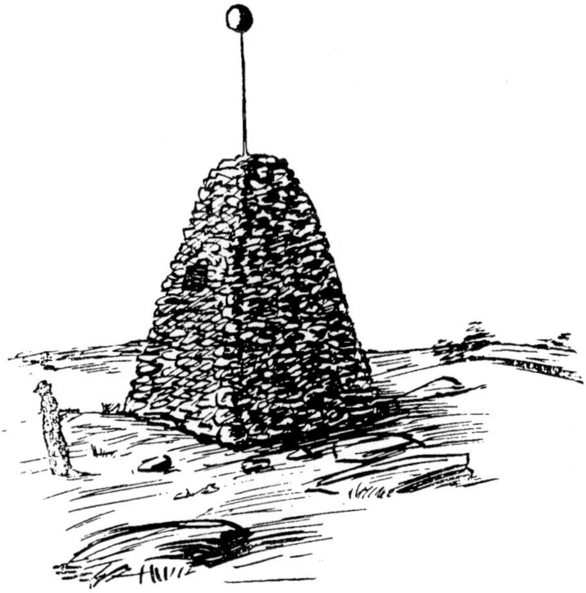

MONUMENT ON BEAR MOUNTAIN.
Erected by Hon. Robbins Bartell, of Norfolk

Housatonic; fertile meadows stretch over much of the country till they meet stately forests and over all is thrown a glamour of history and romance.

Many years ago, before the advent of the white man the Indian loved to make this land his home—his light canoe cut the clear waters of the Wononscopomuc, while by its shores

MAIN STREET SALISBURY.

the council fire gleamed on the visage of the brave warrior, or glared on the face of the tender and helpless.

From Ore Hill in the western part of the town iron ore has been mined for a century and a half, and it was here that a large amount of iron was obtained in the old Revolutionary times for cannon and ball.

RESIDENCE OF MILO RICHARDSON, LIME ROCK.

Lakeville, situated on the sloping shores of the Wononscopomuc is very much before the educational world at present, for it is here that Mrs. Maria Hotchkiss built the Hotchkiss School at a cost of $200,000 and has endowed the school with $200,000 more. Many summer homes have also been built and with their well-kept grounds, add much to the beauty of the village.

Salisbury about a mile a half east of Lakeville, is a typical New England town with a broad street lined with shade trees running its entire length.

The town is most fortunate in possessing the beautiful Scoville Library, one of the finest buildings of its kind in the

LIBRARY.

state. In the library proper over the fireplace is laid a stone taken from Salisbury Cathedral, England, while from the granite tower of the building a four dial clock tells the time of day.

Chapinville, a small hamlet east of Salisbury, has many pretty residences and is well worth a visit.

Probably no prettier or more picturesque hamlet exists in all the county than the little village of Lime Rock, situated in the western part of the town and about three miles from

EPISCOPAL CHURCH, LIME ROCK.

Lakeville. The village lies snugly ensconsed in a narrow valley which is encircled by ragged peaks of the Taconic range. From the precipitous sides of the mountain glittering patches of limestone crop forth, while in contrast are the granite rocks and hemlock trees which clothe a portion of these rock-ribbed hills. In the west part of this little hamlet the noble Housatonic comes rolling down between the hills as it sweeps on toward the sea.

From Butler's bridge which spans the Housatonic, a broad graveled road, lined with graceful elms, extends straight as an arrow for nearly a mile.

The village owes much of its trim appearance to the Barnum and Richardson families, who conduct the extensive

HOTCHKISS SCHOOL.

iron business of the Barnum & Richardson Company, for Salisbury has the reputation of producing the best iron for car wheels in the country.

Back from the road surrounded by velvety stretches of lawn and graceful shrubbery and trees, stand several fine residences from which pretty views may be obtained.

The inhabitants of the town justly feel proud of the pretty stone church which stands at the west end of the street.

RESIDENCE OF CHARLES BARNUM, LIME ROCK.

The township of Salisbury possesses another attraction. Washining and Washinee, "the twin lakes of the woods." These lakes are most accessible, since the P., R. & N. E. Railroad runs along their shores for a long distance. A few months ago the writer stood upon the shores of these beautiful lakes and watched an August sunset. Not a ripple stirred the broad expanse of blue water. In the distance two or three boats were moving toward the white tents which were half hidden by the foliage of the pine trees. Across the

TWIN LAKES.

lake, without a peer in all Connecticut, * Bear Mountain, stood out in bold relief against a dark blue sky which now glowed with all the glories of the departing day.

Slowly the sun, which now looked like a great ball of fire, sank behind Mt. Washington, and the whole landscape was bathed in a subdued light

Let us leave the reader here in the most beautiful spot in all the town, to watch the fine effect of shade and light, till at last day gives place to the beauty and quiet of a mid-summer night.

---

* Bear Mt , the highest point of land in Connecticut, is 2,250 feet above tide water, and is crowned with a stone monument surmounted by a bronze ball. This monument was erected by the late Robbins Battell of Norfolk.

THOMASTON.

Thomaston formerly belonged to the township of Plymouth and is the youngest town in the county.

The town is broken up by hills and is extremely picturesque, for there are deep gorges and wild glens and turbulent mountain brooks. The town lies in the Naugatuck Valley through which the river of the same name flows. In this town

HOSE HOUSE

is the Waterbury Reservoir from which the city of Waterbury obtains its water supply.

Although Thomaston is a manufacturing town it is well laid out, the streets being broad and well lined by fine shade trees.

This town has been made famous by the clock industry, for it is here that the Seth Thomas clocks are made, the yearly output being valued at over a million of dollars.

The largest and handsomest building in the village is the

TOWN HALL.

brick opera house. Near this building is a pretty brick strue-ture known as the Thomaston Hose House.

Picturesque drives radiate from the town leading the tour-ist through a wealth of scenery and beauty.

DAM NEAR KNIFE SHOP.

NEW MILFORD.

This township at one time included the town of Bridgwater and was the largest township in the county.

Although New Milford has lost in area it is one of the most beautiful towns in the state. Dr. Field, who has girdled the globe a number of times says that the Housatonic Valley is one of the most picturesque valleys he has ever visited, and the

INGLESIDE SCHOOL.

township includes one of the most beautiful portions of this beautiful valley, for the scenery in this section is characterized by wildness and grandeur.

It is thought that at one time, thousands of years ago, much of the land in this section was covered by a broad lake, but some mighty convulsion must have occurred and the natural dam (a range of hills to the south) were torn in twain and the broad expanse of water rushed out of the gorge now known as "Lovers' Leap". Although there is great scenic beauty about this "Lovers' Leap" much interest is added to the place by the tradition that an Indian maid, the daughter of a great chief who looked with disfavor on her white suitor, plunged with her lover from the high bluff into the river and met a fearful death.

The village is beautifully located on sloping hills at an elevation of 750 feet and has a perfect drainage.

From the hill near the Rectory a commanding view of hill, valley and dale, may be obtained in all directions, while in the

STREET VIEW.

foreground lies the broad, double street lined with graceful elms. Lovely drives radiate in every direction.

On Aspetuk Hill stands a spacious summer hotel known as the Weautinaug Inn. This building was formerly the mansion of John P. Treadwell, the first proprietor of the famous St. Nicholas, New York.

* The Ingleside School for young ladies is beautifully located in this village, and adds much to the general refinement of the neighborhood.

Through the generosity of Mr. Egbert Marsh the town is enabled to have a fine library. The library, which is English Gothic in style, is built of buff brick with Long Meadow trimmings. Granite steps supported by red stone walls lead to

* The Ingleside School has lately been located in the Weautinaug Inn.

a large vestibule with mosaic tile floor. The building throughout is finished in oak and ash and contains besides the library proper a room designed for the G. A. R. Post.

The building, which has recently been completed, is situated in the central part of the town.

LOVER'S LEAP.

Handsome churches of various denominations are located in different parts of the village.

In fact, New Milford seems to possess all of the necessary requisites for an ideal home.

SHARON.

Although not one of the oldest towns in Co
Sharon possesses far more of the delicate charm o
Colonial days than many of the more ancient townships
able to boast. The reason for this is that its early settle
were not pioneers in the sense of those who leave behind them all the luxuries of life to brave the hardships and terrors of a wilderness. They were, on the contrary, well-to-do citizens,

MUDGE POND.

who, having discovered an easily accessible territory as yet unappropriated, more beautiful and more fertile than the homes their fathers had chosen, came to occupy the new land because it was good, bringing with them their worldly gear, and many of them bringing also that other possession of greater value than mere money can ever be—minds cultivated far beyond the average of persons of their day. Thus their houses were not merely temporary structures which required at a later date to be torn down and replaced, but houses which

SHARON INN.

are good, substantial and fine-looking to-day. Perhaps more than any other one thing, it is the presence of so many of these genuinely Colonial mansions which gives to Sharon the distinctive charm which is conceded by all who know it.

The location of the township is exceedingly beautiful, lying as it does on about the midway terrace of the long hills which, like gigantic garden terraces, slope southward from the

LIBRARY.

Berkshires to form the western highlands separating the romantic Housatonic from the peaceful valley of the winding Webotuck. Here, high enough to command a view of hill and dale, lake and stream, yet sheltered from the worst violence of the easterly winds by the higher hills which lie behind it, the early settlers found an unrivalled situation. Here they could lay out their broad Main street, a mile and a half long with only sufficient curve to prevent monotony. Some of the finest of the old mansions were built on highways leading to this street. Some on highways which are not now in use. Such was the once fine old structure the names of whose builders

STREET VIEW, LOOKING NORTH.

and early owners have not been kept in local remembrance, but which is known as the "Montgomery House" because that general had his head-quarters there for a short time when on his way to join the "Army of the North" in its expedition against Quebec. Few houses of its day surpassed this in location or construction, and in these days of reverence for

SMITH HOUSE.

whatever is historic, or was beautiful before it was old, we can hardly endure to remember that it has been suffered to fall to ruin. A very similar house once stood in Sharon Valley, but of this not even a trace now remains. The Gay homestead,

CLOCK TOWER.

also off the Main street, is a handsome old house still in good preservation, and happily is owned and occupied by the descendants of him who built it. Among the other old and fine houses still in good preservation, is the Sedgwick homestead overlooking a beautiful part of the elm embowered Main street. At the street's southern end, where it is divided into three diverging ways, is another fine old house now occupied as a summer residence by Mrs. Dunlap-Hopkins of New York. At the street's northern end, commanding a view down the park-like Green, is the quaint and charming house built by Mr. Geo. King, and now occupied by one of his family connection. Among the oldest dwellings in the town and certainly one of the' most beautiful, is the many gabled

TIFFANY HOUSE.

house now owned and occupied by the Rev. C. C. Tiffany and wife. It has been greatly enlarged and improved of late years, but fortunately not modernized. The same is true of a charming Mount Vernon-like house owned by Miss Phœbe Perkins. Other old houses that attract admiration and attention from their age, or their fine appearance, or the beauty of their grounds, or all combined, are those of Mr. Charles Sewall, of Miss Lydia Smith, of Mr. Frederick Carter, of Mr. George Goodwin, of Mrs. Chase, of Miss Wheeler, of Mr. W. O. Wheeler, and of Mrs. McClurg. In the grounds of the latter stands one of the very finest elm trees in all New England.

Several of these fine residences were built in "the days when George the Third was King," and at least one of them in the reign of George the Second.

A house of some historic interest is that known as the Gov. Smith homestead. Built before the Revolutionary war by Dr. Simeon Smith, a physician of note in his day as well as an officer in the patriot army. It is one of the finest examples in New England of the architecture of the later Colonial period. It was in the meadow in front of this house that the Hessian

RESIDENCE OF C. ADOLPHE LOW.

prisoners and their guard were encamped on their way down from Saratoga after Burguoyne's surrender, and in this house that the British and American officers were quartered. From Dr. Smith the house descended to his nephew, John Cotton Smith, Governor of Connecticut during the last war with Great Britain, and the last Governor under the charter of Charles the Second. To Gov. Smith's descendants the house now belongs.

Besides its fine views and drives of endless beauty and variety, Sharon has many attractions, the chief of which is undoubtedly its park-like street, beautifully shaded by double rows of arching elms, with a broad central Green, commanded at one end by a tasteful monument to the Sharon citizens who lost their lives in the civil war, and at the other by a very

COOL GATES, RESIDENCE OF ARTHUR DUANE.

handsome clock tower, the gift of the Misses Wheeler to the town. Three well built churches front upon the street, and at its highest point is a very pretty public library building, the gift of Mrs. B. B. Hotchkiss.

Among the newer residences of Sharon is the massive stone house of Mr. Albert C. Perkins, and the handsome modern houses of Mr. C. A. Low and Mr. Arthur Duane.

COOL GATES, INTERIOR OF HALL.

NEW HARTFORD.

New Hartford has charms which should hold the lover of Nature a long time. A thriving village encircled by lofty hills and mountains, and environed by scenery which is unsurpassed in all New England for beauty and picturesqueness.

Far above the town the east and west branches of the Farmington unite forming the river so familiar to every one in this locality.

This river which broadens out in quiet expanses of blue lake in some places, in others goes dashing and eddying over its

RESIDENCE OF CLARA LOUISE KELLOGG.

rocky bed till it comes to a wild wierd spot, about a mile below Pine Meadow, known as Satan's Kingdom. Here the Farmington winds like a silver thread through a deep wild gorge whose rocky and precipitous sides reach upward for several hundred feet till they seem to meet the blue sky. For thousands of years the river has been wearing the deep and narrow gorge through which it flows. Cut into each side of this

NEW HARTFORD, SHOWING A PORTION OF GREENWOOD'S FACTORIES IN THE BACKGROUND.

SATAN'S KINGDOM.

ravine, high above the river, is the solid bed of a railroad over which trains dash, the puffing of the engine echoing and re-echoing on the rocky walls of the cliffs. Far up on the crest of one of the sides of this gorge one may obtain such a view as he will never forget. Here Clara Louise Kellogg, America's Prima Donna has carrolled some of her sweetest melodies, and here Rose Terry Cooke has found an inspiration for some of her sweetest verse. While an author more than twenty years ago, wove a description of this wild romantic pass through the mountain in a story which was published in "Harper's Monthly."

The fact that Clara Louise Kellogg has made this town her home is enough of itself to bring this section before the public eye.

On the east side of the Farmington the Greenwoods Company established their plant in 1812, for the manufacture of duck, and their product has become famous the world over. The writer was fortunate enough to be shown over a large portion of this extensive plant. Passing through a long low fire-proof stone building he saw hundreds of thousands of yards of duck worth thousands of dollars. The product of these mills is shipped to all parts of the world. It was the Greenwoods Company which furnished the duck for the sails for the Defender when she defeated the Valkyrie in the autumn 1895.

New Hartford also has many interesting drives and points which attract the visitor in various parts of the town.

WATERTOWN.

There are few towns in the little Nutmeg state as beautifully located as Watertown, in the western part of Connecticut, and southern part of Litchfield county, and persons who have visited or driven through its valleys and over its hills are ever ready to speak of the picturesque town, with its handsome residences, its productive farms, and the general thrift of its people. Hundreds of years ago, when the whole face of nature was one vast solitude, uncheered by the rays of civilization, when the wild Indian roamed over its solitary hills and valleys, when its forests were in their primitive state, and no sign of cultivation of any kind existed, the imagination can form a vived picture of its grand, romantic, and picturesque appearance. But, if the Indian has ceased to roam over its hills and valleys, if the stately and once gigantic forests have disappeared, it is none the less grand none the less romantic, but more picturesque than it has ever been.

Civilization with its rapid strides, inventions, and improvements has wiped out the old and once familiar landmarks of the old settlers. Now the shrill whistle of the locomotive and the hum of driving industries have broken the solitude of two hundred years ago, and where formerly swamps, forests, and wilderness existed, there are driving farms, happy homes and everything that makes life attractive and worth living.

Every thing has changed for the better. The first permanent settlements of the present town of Watertown were made about 1721, by Jonathan Scott, Sr. and Ebenezer Richardson, the latter one of the descendants of the pioneers of Waterbury. Watertown is bounded as follows: on the north by Morris and Thomaston, on the east by Thomaston and Waterville, on the south by Waterbury and Middlebury, and on the west by Woodbury and Bethlehem. Within a few years, the main part of the town was on a rising hill, but since the railroad has

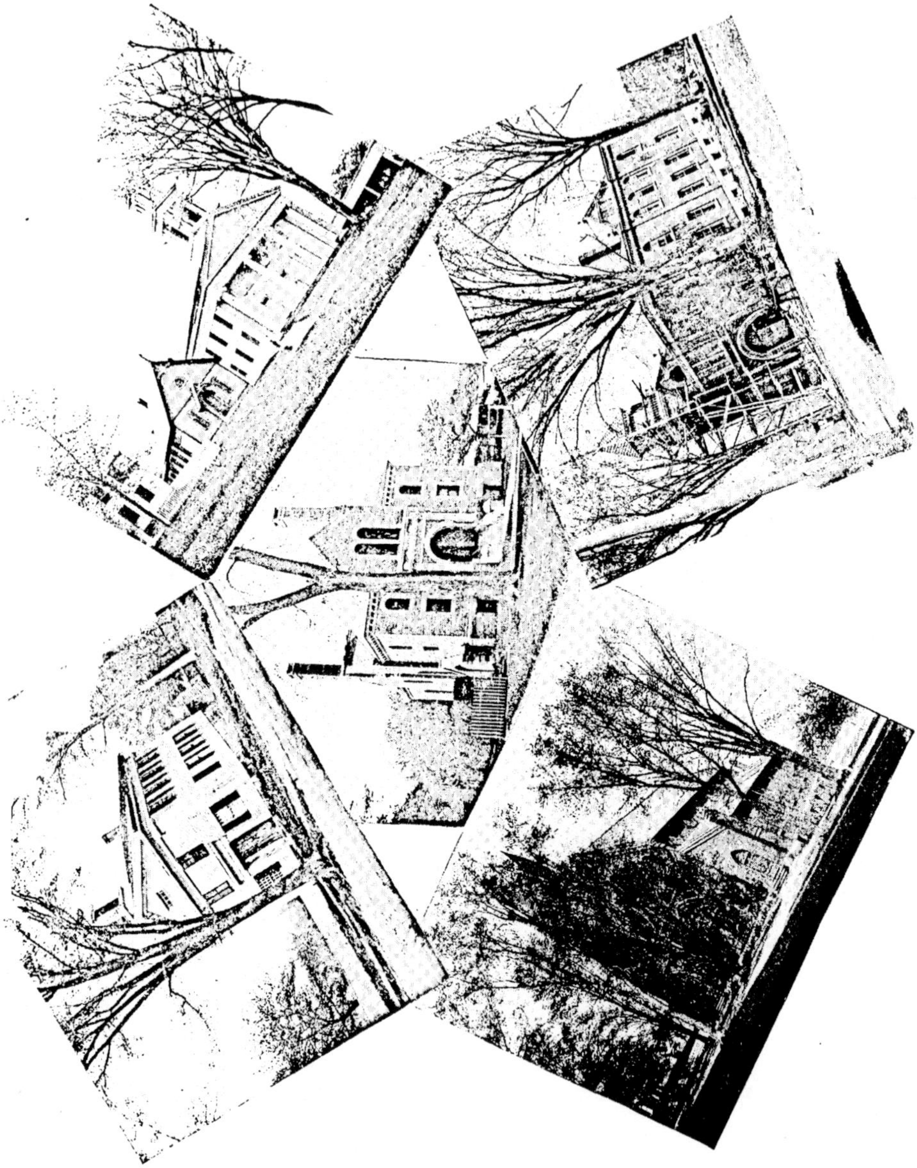

come in use (1870), the town has built up greatly, and its principal places of business are at present under the hill, in the vicinity of the railroad station. From all parts of the compass, but more especially from the south, north, and east the picturesque town with its handsome and modern residences and public buildings, shows to the best advantage. From Nova Scotia hill can be seen Waterbury, Middlebury, and on a clear day the distant Catskill Mountains, N. Y. Watertown was incorporated and became a town by itself June 17, 1780. Prior to that its

STEEL'S BROOK

name was Westbury, and belonged to the present town of Waterbury.

In 1865 the Warren house was erected, and was for many years one of the largest and most prominent summer resorts in New England. Here in Watertown was made the first sewing machine, and here the famous and world-wide known Wheeler & Wilson Sewing Machine Company was organized and carried on a prosperous business for many years. But as the business kept on increasing, and the faculties for shipping were inadequate, the company moved to Bridgeport. In 1866

LIBRARY.

the Watertown Agricultural & Horse Association came into life, and its fairs with its fine thoroughbred cattle became known all over the country. Watertown has two large silk mills where sewing silk is manufactured besides hardware factories. One of the first attractions that meets the traveler's eyes coming from the south is St. John's Roman Catholic church, with its handsome rectory close by, and the beautifully laid out grounds. Opposite the church stands the Pythian hall which was erected in 1892, and is owned by members of the order. As we advance a little further north, we behold the handsome and magnificent residence, the finest in the town, of Merritt Heminway. Opposite this is the old brick store which was erected in 1824. On the rising hill to our right stands the sightly residence of Buell Heminway, which commands a fine view east and south. A little north from here is the Methodist church built 1850. Opposite the church and Mr. Buell Heminway's residence is the green, one of Watertown's beauty spots. Here, westerly stands the Congregational church, built 1840. A little south of this is the Watertown free public library building, built of hewed granite, a gift by the De Forest family. This is the finest structure in our town, and contains nearly 8000 volumes. Opposite this is Watertown's new town hall, with its handsome tower, which was used for the first time May 30, 1895, Memorial day. A little west from here is the most charming residential spot on the hill with its fine green in the center. Here is Christ Episcopal church with its rectory, and spacious parish house. Opposite from Christ church is the hospitable residence of John A. Buckingham in the midst of large and well kept grounds. On the corner stands the castle-like structure, erected by George Woodruff, and now owned by Mrs. Virginia Wood, of N. Y. On the road leading to Litchfield, north of Mr. Buckingham's residence, is the old fashioned, but baronial residence of the W. E. Curtis family, with its large and stately trees, handsome flower gardens, and well kept grounds. A little further west we come again to a green with many large and handsome trees. This is the once famous Warren house, now Mr. Taft's school.

INTERIOR LIBRARY.

Mr. Taft's School was established in 1895, at Pelham Manor, New York, by Mr. Horace D. Taft, a graduate of Yale College in the class of 1883, and for some years a tutor of Latin in that institution. The school began with a staff of three teachers and with ten boarding scholars and seven day scholars.

In 1893 Mr. Taft moved the school to Watertown, where the conditions have proved very favorable, and the school has flourished exceedingly. There is now a staff of six teachers, while the boarding scholars number forty-three and the day scholars sixteen.

The object of the school, as set forth in the school catalogue for 1896 is, "to give boys a thorough preparation for the best colleges and scientific schools, and to make them strong, healthy, and manly men." The school prepares for all colleges and scientific schools, but most of the boys go to Yale. The course of study covers five years, and no boy is admitted who is less than twelve years of age.

Mr. Taft was fortunate in securing for his school the Warren House which, for nearly thirty years was the most famous hotel in this part of the state and which has proved to be a most admirable building for the purposes of the school. The adaptation of the building for these purposes required several changes. A gymnasium and locker room completed the school equipment. The school has the use of an admirable athletic ground in the Watertown fair grounds. The quiet New England village of Watertown with its healthful climate, its freedom from all temptation, and its opportunities for open air exercise is an ideal place for such a school.

West from here on a rising hill, overlooking the town, is the old fair ground, from which a fine view can be obtained. One of the most picturesque features in our town, is Lake Winnimaug, in the south-western part of our town, which covers nearly 2,000 acres, and is considered by all fishermen, as the finest fishing place in Litchfield county. Adjoining the lake, is the well known and productive farm of Charles B. Mattoon, who carries on the largest dairy in the county, keeping no less

TAFT SCHOOL.

than one hundred cows. In the north-western part, are the so-called "Nonnewaug Falls," which are known to all. In the north-eastern part of our town, on the road to Thomaston is, "Black Rock," where a beautiful view of the Naugatuck valley can be obtained. There is an entrance into a cave of this rock which, however, has been little explored. The scenery from the top of "Black rock" is one of the most romantic and picturesque that can be found anywhere.

Watertown is known for its good roads with its fine and shady drives, which are used extensively by people of Waterbury and surrounding towns.

WOODBURY.

The grand old town of Woodbury is not only picturesque but historic. The land is composed of beautiful rolling hills and fertile valleys.

Over two hundred years before the settlement of Woodbury, the Indian loved to make this section his home and here engaged in deadly strife, and the whole land teems with legends

PARKER ACADEMY.

of the brave "sons of the forest." Among those who drew up and signed the first articles by which the settlers of Woodbury agreed to be governed were ancestors of General Sherman, General Grant and Chauncey M. DePew.

Over half of the company which took Fort Ticonderoga were inhabitants of this town.

Woodbury has also taken a leading part in American church history, for in 1783 an event occurred which was not only of importance to the state, but also to North America, for it was at the house of the Rev. Mr. Marshall that a meeting of Episcopal clergymen appointed a clergyman to be ordained as the first Bishop of the Protestant Episcopal Church of America.

Woodbury is not only historic, but it is a beautiful town, for besides the handsome residences the streets are made beautiful with noble elms.

HOUSE IN WHICH BISHOP SEABURY WAS ELECTED.

The Parker Academy, a preparatory school for boys and girls, has already taken a high stand for doing thorough work.

One of the most beautiful portions of the town and one which has many Indian legends is Orenaug Park.

Just back from Main street, and extending parallel with it, rises a ledge of rocks 140 feet high, crowned by stately old pines. This range forms the western front of the beautiful tract of land given to the town in 1889 by Mrs. H. W. Shove, that all might enjoy the privileges of a public park, with the one condition that birds and all animals should remain unmolested. The Park derived its name Orenaug, from the Indians who formally possessed it.

Topographically, the twenty acres are most variable, comprising many delightful ravines whose sloping hillsides are carpeted with ferns and mosses often tenanted by squirrels and partridges. The southern boundary is of stone work 400 feet in length. The entrance on Park Avenue is between

ENTRANCE TO ORENAUG PARK.

monuments, ten feet in height, composed of a great variety of beautifully colored stones, furnishing opportunity for much study in mineralogy. Much work has been done in the Park to bring out the beauty of its natural endowments, which has attracted visitors widely, like pilgrims to a shrine. A fund has been established, giving opportunity to all interested through former residence or association in this grand old town, to contribute, that still more may be done to increase its power of bringing people nearer to Nature. Singing, and Bethel rocks are special places of interest, being associated with many Indian legends, thus adding the charm of romance to their surroundings. Drives are so constructed and openings made through the trees, that one may enjoy the quiet of the dense woods, or feast upon the extended views revealed to one from the edge of the cliff, of the charming valley of the Pomperaug and the "Ancient Town of Woodbury."

ROXBURY.

At one time Roxbury belonged to her sister town, Woodbury. The town possesses much of the picturesque, which, coupled with its historical associations make it one of the most interesting towns in the lower part of the county.

A little more than two hundred years ago the Indian roamed at will over the forest crowned hills of Roxbury, or met in council down by the banks of the picturesque Shepaug, and to this day are found Indian relics near Pulpit Rock, which is situated a short distance north of Roxbury Station.

ROXBURY.

Roxbury is famous as the birthplace of Gen. Ethan Allen. Another of her famous sons was Col. Warner, who figured so prominently in the War of the Revolution, and at Roxbury Center stands a monument erected in his memory.

Although no battle ever took place in the limits of the township the remains of a camp, occupied by Gen. Nixon during one winter in the Revolution, are still pointed out.

It is said that Gen. Washington visited the town at one time, and the story is no doubt true, as the Father of his Country is known to have visited Woodbury more than once.

This town is made up of high and rugged hills from which one may obtain extensive views, but no finer sweep of country may be seen than that which is viewed by one who stands on Good's Hill, which is situated about three miles

MONUMENT IN MEMORY OF COL. SETH WARNER.

east of Roxbury Center. Without doubt the most interesting portion of the town is located along the banks of the Shepaug. Shepaug is a Mohegan name meaning Rock River. Here "Dame Nature" has been most lavish with her gifts, for the scenery is such as might ravish the soul of a poet. Strange and varied are the rock formations in this section while the beauty is enhanced by the silvery river.

West of the river is a locality known as Mine Hill, where are found spathic deposits of iron, chalybeate springs and fine granite quarries, which are owned by the Mower Company.

Following the banks of the winding river for nearly two miles in a southerly direction we come to the famous Shepaug Falls, which have attracted so many tourists. Here we will leave the reader to watch the eddying waters as they flow over the dark shelf of rock and fall in a boiling mass of foam in the caldron like receptacle below.

MINE HILL QUARRY AND CUTTING YARD.

HARWINTON.

Harwinton, which is situated southeast of Torrington, is one of the oldest towns in the county.

Many years before the advent of the white man, the Indian roamed at will over what is now a fine farming district.

The town is composed of rolling hills from which many beautiful views may be obtained.

The town has many historical associations. One story is told how Washington and his troops at one time passed through this place and dined at the old Abijah Catlin place. This house, which stands a short distance from Harwinton center, is in a very good state of preservation.

One of the points of interest is the Huntington Chapel, a neat granite structure, the gift of C. P. Huntington.

ABIJAH CATLIN PLACE.

The pure air and fine views have drawn many people from the large cities to spend their summers in this old town, while a "wild agitation over a proposed trolley" may land hundreds of city dwellers among the everlasting hills of this lovely country village.

HUNTINGTON CHAPEL.

PLYMOUTH.

The town of Plymouth, which lies in the southeastern portion of Litchfield county, was known by the name of Northbury up to 1780. Prior to that time it belonged to Waterbury, but in 1780 was set off with Watertown, and remained so until May 1795, when it was incorporated under the name of Plymouth. The town also embraced what is now

OLD TRAINING GROUND.

Thomaston up till 1875, when the latter place was set off from the town of Plymouth.

The entire town possesses much of the picturesque, and beautiful high hills extend throughout the town, while sparkling streams flow through deep gorges and dash over primeval dams, forming numerous waterfalls. Beautiful drives, consisting in some places of macadam road, built under the supervision of the state, extend in various directions. In some places these roads, broad and smooth, run through growths of woodland, or along the side of some deep wild gorge through which rushes a mountain stream in its hurried course to join some larger stream.

The town is very accessible, as it is within easy reach of two railroads.

Plymouth early became a manufacturing town, and to enumerate the many enterprises that have had their birth in

CIRCULAR DAM.

this town would fill a volume. Some of the greatest inventors of the age have resided in Plymouth, among them being Terry, Thomas, Hoadley and Jerome, whose clocks have gone the world over.

The lock business was first successfully started in Plymouth and has continued to be the life of the town, and no traveler has ever yet gone out of the sound of the celebrated clocks named above, or the click of a Terryville lock.

In 1847 Andrew Terry started the Malleable Iron Foundry, which to-day is an extensive plant situated near the depot.

Not only has Plymouth been foremost in the past in manufacturing and inventions, but in times of war has furnished her full quota. The oldest survivor of the Revolution and last

GORGE, DEVIL'S BACKBONE.

pensioner of that war was Lemuel Cook, who died in 1866, aged 102 years.

The Plymouth Green is also a place of much interest, as it was used for a training ground. At present it is much beautified by graceful elms and also contains a soldiers' monument.

This town was the home of Dorance Atwater, who, while a prisoner in the late war, secured and brought away the only correct list of the Andersonville dead, a list of nearly thirteen thousand. Reader, stop for a moment and think what such a list meant to the friends of the dead and also its worth to the government. It was by this list that Miss Clara Barton and

Mr. Atwater were enabled to direct the making of the graves of those thirteen thousand.

Mr. Atwater was anxious to publish his list, but the government wanted first to make a copy; later Mr. Atwater was refused the list, but he retained it when the work was finished at Andersonville Cemetery and placed it where the government could not find it, claiming it as his personal property secured at the risk of his life and while a prisoner of war. Mr. Atwater was imprisoned for this act, this time in a northern prison.

Through the efforts of Miss Barton before a joint session of the Connecticut Legislature, Mr. Atwater's release was obtained. He was afterward a United States consul and in that capacity served his country faithfully for twenty years.

The famous dead list was exhibited for the first time at Plymouth Centennial, May 15, 1896, at which occasion Miss Clara Barton was present.

Plymouth has been the home of prominent clergymen, skillful physicians and noted members of the bar.

It is not known that George Washington ever slept in any house in Plymouth, but there are now a number of well preserved old houses where he could have secured a night's lodging had he come this way, and left his "Hatchet" at home. Boys in his day did not value young trees, for tradition has it that one of the clergymen of the town set out some elm trees on South street in Plymouth Center, but they were pulled up by the boys, whereupon the old parson announced that if the boys would return the trees as they found them he would give them a gallon of rum. The trees were soon found and put in place, and now a stately row of elms stand as a monument to the good old parson who lived in days when customs differed from those of to-day.

COLEBROOK.

Colebrook lies just west of Norfolk, among beautiful wooded hills, its white church spire rising from the encircling elms and looking down the street of a typical New England country village.

Here are no mills, or railways, or steam whistles, and here a child may wander far and find no possible danger.

Two stores, one containing the post-office, comprise all business done, except farming. The farms are well kept and one sees no evidence of the decadence so often apparent in the hill towns of Connecticut.

Well kept roads diverge from the village and mount the hills, from which lovely views may be obtained in every direction, or wind down the village beside some picturesque trout stream.

Colebrook is remarkable for the beauty of its roadside verdure.

> "Boon Nature scatters free and wild
> Each plant and flower, the Mountain's Child."

Ferns grow in endless variety and luxuriance, clematis and wild grape vines festoon the shrubbery, and all kinds of wild flowers are found each in its season including many varieties rare or unknown in most localities, making the town a delight to the botanist as well as to the lover of beauty.

A favorite drive of ten miles and return follows Sandy Brook, deep among the wooded hills, to Riverton, and there follows the Farmington on one bank as far as Pleasant Valley returning on the other side of the river.

When near Robertsville, a small hamlet in the southern part of the township, one should not fail to visit the celebrated Falls of the Tunxis.

COLEBROOK.

TUNXIS FALLS.

Here "Dame Nature" has scattered her charms in prodigal profusion, the result being that this spot has been the Mecca of thousands of tourists and jolly picnicers. Here the sparkling waters of the Tunxis plunge down a rocky defile till it is lost in a deep pool at the foot of the dark rock.

> "O Tunxis! would I had a poet's powers
> To wreath thy beauties for eternity."

This was the Tunxis of a few years ago, but the onward march of this modern age has now invaded this beautiful spot. Dynamite has reduced the picturesque rocks into a thousand broken fragments, and in their place stands a power house, while the rushing flood has been harnessed and furnishes electric lights for the busy town of Winsted four miles distant.

Another pleasant drive is to Highland Lake in Winsted five miles south. To the north lie the hills of Sandisfield from which extensive views include the Taconics, the Catskills, and mountains of Berkshire. In every direction is some interesting objective point for a driving excursion.

Several years ago Colebrook was discovered as a place of summer sojourn for those in search of genuine seclusion and rural beauty and is growing in popularity year by year.

Although Colebrook was the last settled town in the county, yet many of her sons have been prominent in affairs of the nation and the little town located in the hills has no need to feel ashamed of her past record.

TUNXIS FALLS.

BARKHAMSTED.

Barkhamsted possesses much of the picturesque, for it is composed of rugged and varied areas through which flow mountain streams that have been used as motive power for many manufacturing firms.

CONGREGATIONAL CHURCH.

The town is divided into three sections, Barkhamsted Center, Pleasant Valley, and Riverton.

In the first of these districts is located the Town Hall, a pretty church, and a small cluster of houses.

Between the two last named districts are wild and romantic gorges and many pretty bits of scenery.

The principal part of the business in town is transacted at Riverton, a small village located about four miles from Winsted.

This little town has been famous for years for the large number of fine rules which have been manufactured here.

The streets are broad and well shaded by fine old trees, while the west branch of the Farmington, which flows directly through the center of the town, adds much to the beauty and picturesqueness of the place.

BETHLEHEM.

Bethlehem, although a small town is very interesting, as it has many historical associations.

The occupation of its inhabitants is chiefly agriculture, for there are many fine farms on these rolling hills.

Among the early settlers was Hezekiah Hooker, a descendant of the celebrated Rev. Thomas Hooker of Hartford, and a man who did much for the town during its early existence.

This town, which lies north of Woodbury, was much frequented when first settled, by the Indians, who are said to have built their wigwams on the banks of the stream just below the celebrated Nonnewag Falls.

These falls are situated in the southeastern part of the town in a romantic glen and form a beautiful cascade.

Bethlehem has at different times had schools within her borders which have drawn many scholars who have become famous in after life.

It was at one of these schools that the famous South Carolina Statesman, John C. Calhoun, spent a portion of his boyhood.

With all her delightful qualities, Bethlehem should be known to the outside world at no very distant date.

CORNWALL.

Litchfield County has been termed a veritable Switzerland, but for wilderness, picturesqueness and grandeur, Cornwall is unsurpassed by any of her sister towns. A series of little villages, perhaps we might call them hamlets, lie ensconced among the rugged peaks of the Cornwall Range and have all the charm and seclusion of some little Alpine town.

From the high eminences extensive views of Litchfield and Berkshire Hills are seen, while to the southward may be

RESIDENCE OF HON. T. S. GOLD

obtained glimpses of Long Island Sound. It is in the southern part of the town that the celebrated tower (now in ruins), crowns Mohawk.

The Housatonic River forms the western boundary of the township, adding much to the beauty of the place.

These wild and picturesque Cornwall neighborhoods have furnished many a subject for the painters brush.

HOUSATONIC VALLEY INSTITUTE.

Cornwall is one of the most highly developed farming districts of the state. It is at Cream Hill that the Hon. T. S. Gold, Secretary of the State Board of Agriculture, has his home.

Near the Cream Hill farm lies a beautiful sheet of water which would compare favorably with a New Hampshire mountain lake.

The village known as Cornwall Plains is one of the prettiest in this section, for the land is very level while a short distance away, steep mountains rise from the plain and add to the

RESIDENCE OF JOHN E. CALHOUN.

beauty of the scenery. Among many pretty homes may be mentioned the fine old Colonial house owned by Mr. John Calhoun, known as " Coltsfoot Farm.

The Housatonic Valley Institute, a model school, is also located at this village.

A short distance south of the village is located one of the finest groves of white pine in the state, the property of Mr. Calhoun.

The visitor should not fail to drive from Cornwall Plains to Cornwall Bridge for the road is not only a fine carriage drive but the quality and grandeur of the scenery is enhanced with every mile of progress.

The historical associations of the town of Cornwall are of large interest for transactions to the extent of "affecting the weal or woe of the world have hereabouts taken place." It was within this township that the American Board of Commissioners for Foreign Missions, established the first school for the education of young people from foreign lands.

It is in Cornwall that General Sedgwick, who is known to every school boy lived and died.

The town also has many other interesting features which space will not permit us to mention.

KENT.

Many years ago this town belonged to Warren, but as it stands to-day is bounded on the north by Cornwall, east by Warren and Lake Waramaug, south by New Milford and west by the towns of Dover and Amenia in New York state.

The whole township is broken up by mountains from 800 to 1,400 feet high, while deep valleys lie between. The mountain sides in many places are rocky and precipitous; in others the rise is more gradual, the sides of the hills being covered with fine farms.

The town is one of the most picturesque in the county, for besides the mountains there are several small lakes, while in the southeastern portion of the town lies a part of the beautiful Lake Waramaug. The Housatonic River runs through the entire length of the town. Every mile of this nôble river abounds in great scenic beauty, for in places it flows through deep and narrow gorges, over huge boulders and rock masses and finally widens out into clear expanses.

The lakes and various mountain streams abound in fish, and for this reason alone have become famous. A natural gap in the West Mountains gives easy access to South Dover on the Harlem Railroad, a distance of 69 miles from New York.

Four miles south from Kent Station the Housatonic River breaks through a limestone ridge and tumbles over the rocks, falling fifty feet in leaps and whirls, affording a beautiful sight, while at high water the view is grand and imposing.

Four miles north from Kent Village a mountain stream, running through a ravine densely shaded by hemlocks, falls a distance of nearly 400 feet in a succession of cataracts and rapids. Tourists pronounce these falls finer than the celebrated Minnehaha Falls.

Very many pretty drives radiate from the village and wind over mountains and down through fertile valleys. From some points on the Kent hills on a clear day may be obtained

KENT FALLS.

fine views of the Catskills, while the white sails of vessels may be discerned on Long Island Sound.

West of the Housatonic are the old tribal grounds of the Seate Cook Indians, on which a remnant of the tribe still live.

Although the village is small and the chief business of Kent is farming, the town has "all the dignity of an eminently respectable New England village" enhanced by scenic beauty which will compare favorably with the White Mountains.

ROADWAY NEAR GREAT BEND. GREAT BEND IN THE HOUSATONIC. STREET VIEW.

GOSHEN.

The township of Goshen not only possesses much natural beauty, but has many interesting features and points of interest. The fact that it ranks in elevation among the highest in the state insures sweeping views and pure air.

IVY MOUNTAIN TOWER.

Three ridges of land run from north to south with valleys between and make the landscape one of unusual diversity, while the Green Mountains terminate within the borders of the township.

The town is divided into sections known as, Goshen Center, West Goshen, North Goshen and Hall Meadow located in the eastern part of the town. The last named section was bought by Asaph Hall when the town was first settled. Mr. Hall was the grandfather of Prof. Asaph Hall, of Washington, D. C., who has recently built himself a home among the pleasant hills of his native town. The road known as East Street, was once the prominent route in Goshen, it being a direct way from Norfolk to Litchfield. The view from this

LOG CABIN. PROF. HALL'S RESIDENCE.
TIPPING ROCK.

street is wonderful not only showing a variety of scenery, but of wide extent. To the west the Catskills are visible while the spires of churches in neighboring towns may be seen from the east, with Talcott Mountain beyond.

In 1884, a tower seventy feet high was erected on Ivy Mountain by Franklin Wadhams and from no other place in this section can so fine a view be obtained. It is said on a clear day one may see the waters of Long Island Sound, while the sparkling waters of several lakes, which are scattered throughout the town, may be seen.

The town possesses a most wonderful natural curiosity in the north part of the town, which is the Mecca of hundreds of people who spend their summers in Litchfield and surrounding towns, known as Tipping Rock.

Thousands of years ago when mighty glaciers swept over our hills, this rock was borne from some northern locality and left in the position it now occupies a wonder to all who view it. The ground in the vicinity of this rock is covered with many boulders of fantastic shape and size. Geologists tell us (from the many scars and scratches on the flat rock on which the boulder rests) that the movement of the rock was from northeast to southwest. The rock which is of granite formation is estimated to weigh ninety tons and is so balanced that a child can sway it back and forth, the swing being nearly six inches. The boulder measures nine feet nine inches in height, thirteen feet three inches in diameter and forty feet in circumference.

Goshen has obtained a reputation as a fine farming country and her products, such as cheese and butter, command high prices in the city markets. But Goshen has many requisites for an ideal summer resort, already a few are taking advantage of the cheap price of land while the fact that the town will, in the near future, undoubtedly have a trolly line which will make it accessible to the outside world, will be a means of dotting the hill tops with summer residences from which may be obtained pretty views of brook and forest, hill and vale, which render the town beautiful to the lover of Nature.

The town of Canaan, which includes the villages of Falls Village and Huntsville, possesses much to attract the lover of beauty.

The surface of this town is more mountainous and rugged than its sister town, North Canaan, and the chief occupation is agriculture. But he who admires the beautiful will find a

HUNT SCHOOL.

pleasing diversity of mountain, valley, and wood, while the broad and romantic Housatonic comes rolling down between the rock-ribbed hills till near Falls Village it plunges over a broad shelf of rock forming the far-famed "Falls," which has been a subject for artists and photographers for many a year.

These falls are said to furnish one of the finest water powers in the "Nutmeg" state, and its close proximity to the Berkshire division of the New York, New Haven and Hartford Railroad should make an ideal site for some enterprising manufacturer. The visitor to this quiet town should not forget to saunter up the quaint old street, lined with wide-spreading trees, which forms the main thoroughfare of the town. On both sides of the street are old-fashioned dwellings

HOUSATONIC FALLS.

and business blocks, but even here modern architecture appears in the handsome new "Iron Bank" which was erected a few years since.

Although nearly every town in the county has a good secondary school, Canaan is able to hold her own in educational matters, for seven years ago in accordance with bequests of the late Wealthy A. and Catharine E. Hunt, a school and library were founded in memory of their late brother, David M. Hunt. The school building, built of brick, and consisting of two stories, occupies a pleasant site in well kept grounds. The upper story of this building contains the David M. Hunt Library. The school is under the charge of a competent principal and corps of teachers. In various parts of the town are many fine views which are quite accessible and which prove attractive to the visitor

LAKE WARAMAUG.

WARREN.

It has been said that more ministers have come from Warren than any other town in the county. Perhaps it may be from the fact that the little town is set high on a hill above all of the surrounding towns and is removed from the contaminating influences of the outside world. Although the town is somewhat isolated from other communities, it has many interesting features and a charm and beauty for the lover of nature.

On the farm of Justice Comstock is located one of the finest granite quarries in the state, while the farmhouse which is a hundred years old, is built of material taken from this same quarry.

A short distance southwest of the village is located a hill, known as Above-all, from whose summit sweeping views of hill, valley, and lake may be seen; while on a clear day the blue waters of Long Island Sound may be seen in the distance.

But he who would see the most beautiful portion of Warren, should visit Lake Waramaug. This lake, which is situated in the townships of Kent, Washington, and Warren, has the greater part of its area in Warren. This beautiful sheet of water is one of the largest and most picturesque in the state, being surrounded by high bluffs and hills whose rocky faces reach to the waters edge. A good carriage drive has been built around this lake, a distance of ten miles, and in the whole drive there is no bit of sameness, for in many places it almost touches the limpid waters, or lies at the foot of bold bluffs and crags while the greater portion passes through a fine growth of woods.

From many of the commanding sites which surround the lake handsome summer residences and large hotels have been built, for nearly five hundred people summer in this sylvan spot.

From the Warren shore the finest view may be obtained from the hill on which the Hopkins place is built. Truly within the borders of Warren is located one of the most beautiful spots in all Connecticut.

MORRIS.

Beautiful for situation is Morris, a little town six miles by three, nestling among the hills to the south of Litchfield. Lovely beyond the power of words to describe are its sunny hills and sheltered valleys, its broad lake shimmering in the sunlight, and its peaceful farmhouses looking out from the surrounding trees. Seen on a sunny May day when the trees have on their bridal array, or when the October sun has painted the woods in scarlet and golden its picturesque beauty will not suffer by a comparison with the famous Berkshire hills.

To quote from an old geography "The surface is diversified by hills, mountains and valleys. The soil, is better adapted to grazing than tillage, but this very lack of adaptability for agricultural purposes only adds to the picturesque beauty of its scenery. The hills lie in long undulating ridges stretching from north to south, higher and more precipitous in the western part of the town, where two are of such height as to gain the title of Mt. Tom, and Little Mt. Tom.

Mt. Tom is a cone shaped wooded knoll at whose base on the northwest, lies Mt. Tom pond, a small sheet of water but deep and beautifully clear. A tower crowns the mountain's summit.

From this tower one gets an extensive view of the surrounding country, Litchfield, Bantam and Romford villages—Bantam Lake not far off, and on a very clear day a glimpse of the Catskills, dim and blue in the distance.

But the most picturesque views in the town are around Bantam Lake. Persons who have travelled far and wide have assured me that the view from Whittlesey's hill near the southeast shore of Bantam Lake is one of the finest in the country.

This lake is three miles from north to south and from three quarters of a mile to one and three quarters miles wide. There are numerous summer resorts around it from the plain cottage which almost deserves the name of "shanty" where parties may camp out and "rough it," to their hearts content,—up to the stylish villa, the Club House, and the Dutch Colonial Hotel where one can have city luxuries and prices.

Not far from the southern point of the lake is Camp Columbia, a summer school connected with Columbia College. A farm was leased for a term of years, buildings remodeled, and arranged for dining-halls, kitchen, offices, etc., tents pitched for sleeping accommodations and here about 100 students are taught surveying and map-drawing in the most practical manner.

Although Morris is now a quiet little town, only enlivened during the summer months by city guests attracted hither by the pure air and lovely scenery,—it can boast of a fame and renown, "in the days long since gone by," when Morris Academy was in the zenith of its glory. This famous academy was established in 1790 by General James Morris, the greater part of whose life was spent in this place then called South Farms. While the academy was under his care there was an attendance of nearly 1500 different pupils from twelve different states and from the islands of St. Thomas and Bermudas. More than sixty of his pupils entered college many among them afterward became men of note, among whom were Rev. Edward Beecher, Rev. John Pierpont, and John Brown of Ossawatomic of whom the poet Whittier says:

"John Brown of Ossawatomic they led him out to die,
And lo! a poor slave mother with her little child pressed nigh.
Then the bold blue eyes grew tender and the harsh old face grew mild
As he stooped between the jeering ranks and kissed the negro's child.
The shadows of his stormy life that moment fell apart,
And they who blamed the bloody hand forgave the loving heart.
That kiss from all its guilty means, redeemed the good intent,
And round the grisly fighter's hair, the martyr's aureole bent."

BANTAM LAKE.

The old building, a plain bare wooden structure after the fashion of a century ago, remained standing in a dilapidated condition until about ten years ago.

Another place of historic interest is the birth-place of John Pierpont, poet, preacher and temperance orator. Every school-child has read or learned "Warren's address at the Battle of Bunker Hill," of which he is the author. His old homestead which he describes as

> "A modest mansion in a sunny nook
> Tall trees behind it and a babbling brook
> Flowing in front"—

is situated in the southeast part of the town on the direct road to Thomaston,—is a wide, low farmhouse, which although repaired and modernized to some extent, still retains the same quaint shape and appearance which characterized it during the boyhood of the poet, almost a hundred years ago.

There are many houses in town which date back to Revolutionary times although their number is diminishing yearly. There are many lovely little nooks which are a delight to the artist, and which have formed the subject of a photograph, a sketch or an oil study.

BRIDGEWATER.

The township of Bridgewater is a characteristic New England rural district, which at one time belonged to New Milford.

The land is hilly and agriculture is the leading pursuit.

The town, like several others in Western Connecticut possesses a most beautiful feature, namely the Housatonic River which affords much variety to the scenery.

Quiet and beautiful is this little town with its roadsides lined with wide-spreading trees, while fertile farming districts stretch out toward New Milford.

Churches of several denominations are found in the town and the inhabitants are intelligent and well-to-do.

WASHINGTON.

The traveler arriving by the winding Shepaug Railroad, is quite prepared to appreciate the succession of delightful and peaceful views which are opened up during the ascent of nearly a mile of good road, leading to the rural village of "Washington Green." The town was named Judea, by the first settlers. The only palliation for the change, is that the immortal George is said to have made *this* his first namesake.

It has been said that the church was the Queen Bee of the Puritan community, so it is not surprising to find that as early as 1742, twenty-six members of Judea society unanimously and lovingly agreed upon a place for a meeting-house. This was the site it now occupies.

In 1748 the Reverend Daniel Brinsmade became the pastor and continued in the office until 1793. Like most of the settled ministers of that day, he acquired a large tract of land, and his family took root here to such an extent, that with one exception all his descendants are still residents of the town. Naturally they have had much to do in shaping the character of the town in all directions.

It was a magnet (now Mrs. Gunn) in the Brinsmade family which fastened Mr. Gunn and led to the founding of the Gunnery, a boarding school for boys in 1850. Mr. Gunn, always loyal to the best interests of his native town, made the terms for day pupils such, that all the young men and maidens could secure the advantages of a permanent school of high order. The home of the Gunnery was so attractive and the boys so happy in the town that many have returned to make this the summer home of their mature years.

William Hamilton Gibson owed much of the bending of the twig which became the beautiful tree of knowledge of nature, animate and inanimate, to Mr. Gunn and his free life in Washington. He came here for his summerings many years.

Here he made the first exhibition of his wonderful charts in his never to be forgotten morning talks. His beautiful home, the "Sumach's," was hardly completed when suddenly death removed him in 1896. His life conferred distinction upon the town and was a benediction as well.

E. K. Rossiter the architect, also a Gunnery boy, has an interesting and charming place near the Green. It includes the Rocking Boulder called "Sampson's Rock." The house is built upon a crag, and the place is fitly called "The Rocks." Mr. Rossiter has done more than any other man to open to

WM. HAMILTON GIBSON HOUSE.

view the heart of Picturesque Washington. About ten years ago he became owner of Steep Rock, a remarkable ledge five hundred feet high, overhanging the Shepaug, which runs in the form of a horse-shoe at its base. By degrees he has added to this four hundred acres of what may be called primeval forest. Here Mr Rossiter has built about six miles of driving road following the winding of the river, which it crosses several times by low bridges, over bluffs and ridges under the Rock, and on to its summit. This drive is pronounced by intelligent travelers unsurpassed in loveliness in this country or any other. To get its ravishing beauty one should take this drive in the time of the Laurel bloom. This shrub is here in great abundance. Much of it is of primitive growth reaching the height of fifteen feet. Scattered on the hillsides among the hemlocks or bordering the lower roads, the effect of the dainty pink clusters of this mountain flower is fine. The moral beauty of this gift to the people exceeds the material. Generations yet unborn will bless Mr. Rossiter for this, as well as for the hundreds of shade-trees he has planted by the roadsides.

RESIDENCE OF MR. E. K. ROSSITER.

A cluster of three beautiful cottages, with extensive grounds, occupies the western slope of the hill commanding an extended view of the valley and distant hills; they are owned by R. S. Barnes (another Gunnery boy) and the families of Mairs and Lecome. "The Knoll" which now includes the "Grove" is the delightful and artistic home of E. H. Van Trigen of New York, the pioneer of the cottagers here. In close proximity to this is the Gunnery, which in its forty-seventh year, sustains its excellent reputation under the care of Mr. and Mrs. J. C. Brinsmade—its extensive athletic grounds include the Gymnasium, and connects with "The

Ridge" an attractive Old Colonial home of recent date where Mr. W. G. Brinsmade has a home school for a limited number of boys.

In this neighborhood for nine months of the year most of the real life of the town centres, coasting, skating, foot-ball, dancing and other evening amusements keep the active young blood from stagnation.

The walks and drives about Washington are numerous and varied in interest. Between breakfast and dinner, one can take the fine level road following Bee Brook, with its many turns, its rustic bridges to Lake Waramaug, back over Baldwin

THE GUNNERY.

Hill, and feel that nineteen miles of more beautiful scenery would be hard to find. A hilly road of seven miles in another direction reaches Bantam Lake, the largest in the state. The Walker Brook road, on to Church Hill down through the woods, by the brook-side to Judd's Bridge and home by the cool and shady River-road. Mt. Tom with its pretty lake and the tower on the summit, back by "Chittenden's four corners" (now Woodville) following the river to Romford are drives full of interest and beauty. Painter's Hill should be taken in the afternoon to see the sunset behind The Catskills, The

Hemlocks, Whittlesey Valley, Bell Hill, Nettleton Hollow, and enough others to allow a new drive every day for six weeks. Then for a morning walk, or evening stroll, what can be more beautiful than Sunset Hill (where a vesper service is held quite often on Sunday evening in summer) unless it be the Old North road to Malbury Brook or the "Overlook road" from which can be seen the whole village with the distant hills

ATHLETIC GROUND, GUNNERY.

as a background. Social life in Washington is at all seasons comfortable, but not conventional—the habits and manners are rural, but not rustic. Two well sustained churches on the Green, a free reading-room with library of nearly three thousand volumes founded in 1854, an Entertainment Hall where lectures, concerts, dramas etc., are often given through the year, furnish opportunity for amusement and improvement.

For those so disposed tennis and golf give abundant chance for active exercise.

Mr. Frank Heath in the summer of '96, with generous hospitality laid out an exceptionally fine golf course of nine holes on his own land. This commands one of the finest views in the town and has many national hazards, as stone walls, ditches, trees, brooks etc. There is also an excellent spring of cold water on the place.

ROAD TO STEEP ROCK.

Fishing for black bass in the lake and trout in the brook affords much sport in the season—success is capricious—the same can be said of hunting.

Mr. Arthur D. Woodruff has a pleasant summer home here, "Orchard Terrace" and contributes to the summer pleasure by his own music, and that of his musical friends who visit him. There is no hotel in Washington but several excellent boarding-houses, where many deligntful people find pleasant quarters during the summer months, among these are found preachers, doctors, bankers, lawyers, artists, musicians and many of lesser note. Flora McFlimsey and Mrs. Grundy never come,

BRIDGE TO LUCKY ISLAND, STEEP ROCK PROPERTY.

VIEW ON SHEPAUG FROM STEEP ROCK.

UNDER STEEP ROCK.

"Valley Station" is the attractive stopping-place of most of the cottagers who find a shady drive free from dust, to their homes. Within a stones throw is the "Holiday House" built by E. H. Van Ingen and given to the "Working Girls' Club" of St. Bartholomew's church New York; the house and all its appointments are first-class, it accommodates about sixty and is generously and pleasantly conducted.

"Washington Depot" one mile from the Green is a neat and thrifty village containing four churches and about sixty houses. Here all the business of the town is done. Nearly all

BONFIRE IN GROVE NEAR GUNNERY.

kinds of merchandise is to be found here so satisfactory as to supply most of the demands of citizens and summer people.

The farms are well-kept and productive, mostly occupied by their owners. The farm-houses are neat and tasteful.

Another point of interest in Washington is the quaint old town of New Preston, situated in the northern part of the township and near the shores of the beautiful lake Waramaug.

New Preston has several quaint old houses and there are many historic associations connected with the place. In this town is situated the Upson Seminary, a school for boys, which has already obtained a wide reputation.

New Preston's natural beauties are unexceled by few towns. From the Pinnacle, which is a rugged height of land rising from the waters of Lake Waramaug may be obtained a view which will reward him who makes the steep climb. To the west and north are a "wild tumult of hills" and misty valleys, to the south lies the picturesque Shepaug valley while at our feet shimmers the blue waters of the lake.

UPSON SEMINARY.

Wood, hill and dale vie with each other in all directions to make up a beautiful and varied landscape

The following we quote from the "Life and letters of Dr. Bushnell, who was born in New Preston." The outlet of Lake Waramaug is from the southern end and pours its foaming stream through a narrow valley from which hills on either side rise steeply. The mills and shops that line this stream and use its water power, and rugged farms that climb the hillsides compose the village of New Preston, that still nestled in the safe seclusion of woods and mountains, keeps much of its old character of isolation from the world.

Spotted something that isn't quite right?

Let us know and we'll fix it for future reprints.
And as a thank you for your diligence, we'll
send you a free Sothis book of your choice.

publisher@sothis.press

This page intentionally left blank

Except of course for that line above. And this one.
And whilst we have your attention, perhaps you'd
like to follow us on Instagram @sothis.press. Go on.

Made in United States
North Haven, CT
05 November 2026

21817320R00088